FUNDAMENTALS OF INTERNATIONAL OCCUPATIONAL HEALTH AND SAFETY LAW

BY

PETER C BARNETT-SCHUSTER

This book is printed on acid-free paper.

Published by Aberdeen University Press Services.

Printed in the United States of America
ISBN: 978-0-6152-1490-0

This book is printed on 6" x 9", perfect binding, 60# cream interior paper, black and white interior ink, 100# white exterior paper, full-color exterior ink. Prices are subject to change.

Cover Title Designed by Aberdeen University Press Services.

FUNDAMENTALS OF INTERNATIONAL HEALTH AND SAFETY LAW

First Edition

Peter C Barnett-Schuster

"This book is dedicated to all the construction operatives around the world who take un-necessary risks within their construction workplaces because of the lack of awareness for health and safety law and pressures from their superiors to just get the job done."

TABLE OF CONTENTS

Chapter 1

WHAT DO WE MEAN BY OCCUPATIONAL SAFETY AND HEALTH?

Occupational safety and health (OSH) is a multi-discipline issue concentrated with the protection for safety, health and welfare of people engaged in work or employment. As a secondary effect, Occupational safety and health (OSH) also aims to protect co-workers, family members, employers, customers, suppliers, nearby communities and other members of the public who are affected by the workplace environment.

Since 1950, the International Labour Organisation (ILO) and the World Health Organisation (WHO) have agreed a common definition for occupational health. It was adopted by the joint ILO/WHO Committee on Occupational Health at its first ever meeting in 1950 and revised it at its twelfth meeting in 1995.

The definition reads: *"Occupational health should aim at: the promotion and maintenance of the highest degree of physical, mental and social well-being of workers in all occupations; the prevention amongst workers of departures from health caused by their working conditions; the protection of workers in their employment from risks resulting from factors adverse to health; the placing and maintenance of the worker in an occupational environment adapted to the physiological and psychological capabilities; and, to summarise, the adaption of work to man and of each man to his job."*

What is Health?

The term *'health'* when referring to work indicates:

Absence of disease and infirmity

It indicates the physical and mental elements affecting health – directly related to safety and hygiene at work

Health hazards fall into four categories

1. *Physical*
2. *Chemical*
3. *Biological or*
4. *Carcinogenic*

What do we mean by safety?

A word used in all countries but it's very difficult to find a consistent definition.

Some definitions say it is an *"absence of danger"* but the same dictionary says that danger is *"the absence of safety"*

Some state *"a state of protection or a condition not involving risks"*

There is no definite state of absolute safety, as there is always a chance that a risk of something going wrong, despite how small that chance may be of actually occurring.

Chapter 2

WHY DO WE NEED TO HAVE STANDARDS?

Setting good occupational safety and health standards are agreed to be needed in most countries for the following reasons:

Moral – Employees should not need to risk injury during the

carrying out of work and neither should anyone else connected, associated or coming in contact with the work environment.

Economic – Many governments worldwide understand poor occupational safety and health performance tends to result in cost to the state, region or country. This could be through payments via a welfare scheme i.e. social security payments to the incapacitated, costs for medical treatment and the loss of the *"employability"* of the worker.

The employer will also incur costs in the event of an incident at work occurring such as legal fees, fines issued by the authorities, compensatory damages by the injured parties involved, accident investigation time, possible lost production time, lost goodwill from the workforce, from customers and from the wider community.

Legal – Occupational safety and health requirements may be reinforced in civil law courts and or criminal law courts by means of prosecution; it is accepted that without the extra *"encouragement"* of regulatory authorities taking action or litigation, many organisations would not comply with their moral obligations.

Chapter 3

IMPLEMENTING LEGISLATION

Different countries take varying approaches to Occupational Safety and Health legislation, regulation and enforcement.

European Union (EU), member states have enforcing authorities to ensure that the basic legal requirements for occupational safety and health are complied with. In many European Union countries there is strong cooperation between employer and worker organisations for example Unions, to ensure good Occupational Safety and Health performance as it is recognised it has benefits for both the worker (through maintenance of health) and the employing company (through improved productivity and quality). In 1996 the European Agency for Safety and Health at Work was founded.

Member states of the European Union have all included into their national legislation a series of directives that help set minimum standards of occupational safety and health. These directives of which there are about 20 on various topics follow a similar arrangement requiring the employer to assess the workplace risks and put in place preventative measures based on a hierarchy of control which starts with elimination of the hazard and ends with personal protective equipment.

United Kingdom (UK) health and safety legislation is drawn up and enforced by the Health and Safety Executive and local authorities or local council under the Health and Safety at Work etc. Act etc. 1974. More than ever now in the UK, the regulatory trend is away from prescriptive rules, and towards risk assessment. Recent changes to the laws governing asbestos and fire safety management embrace the concept of risk assessment.

United States of America (USA) the Occupational Safety and Health Act of 1970 created both the National Institute of Occupational Safety and Health (NIOSH) and the Occupational Safety and Health Administration (OSHA). OSHA, in the U. S. Department of Labour, is responsible for developing and enforcing workplace safety and health regulations. NIOSH, in the U.S. Department of Health and Human Services, is responsible for research, information, education and training in occupational safety and health.

OSHA has been regulating occupational safety and health since 1971. Occupational safety and health regulations of a limited number of specifically defined industries was in place for several decades before that and broad regulations by some individual states was in place for many years prior to the establishment of OSHA.

Canadian workers are covered by provincial or federal labour codes depending on the industry they work in. Workers covered by federal legislation including those in mining, transportation and federal employment are covered by the Canada Labour Code. All other workers are covered by the health and safety legislation of the province they work in. The Canadian Centre for Occupational Health and Safety (CCOHS), an agency of the Government of Canada, was created in 1978 by an Act of Parliament.

The act was based on the belief that all Canadians had "...a fundamental right to a healthy and safe working environment." CCOHS is mandated to promote safe and healthy workplaces to help prevent work-related injuries and illnesses.

In **Malaysia**, the Department of Occupational Safety and Health (DOSH) under the Ministry of Human Resources is responsible to ensure that the safety, health and welfare of workers in both the public and private sector are upheld. DOSH is responsible to enforce the Factory and Machinery Act 1969 and the Occupational Safety and Health Act 1994.

Chapter 4

PROBLEMS WORLDWIDE

In general, health and safety in the workplace has improved in most industrialised countries over the past 20 to 30 years. However, the situation in developing countries is relatively unclear largely because of inadequate accident and disease recognition, record-keeping and reporting mechanisms.

It is estimated that at least 250 million occupational accidents occur every year worldwide. 335,000 of these accidents are fatal (result in death).

Since many countries do not have accurate record-keeping and reporting mechanisms, it can be assumed that the real figures are much higher than this.

The number of fatal accidents is much higher in developing countries than in industrialised ones. This difference is primarily due to better health and safety programmes, improved first-aid and medical facilities in the industrialised countries and the active participation of workers in the decision-making process on health and safety issues. Some of the industries with the highest risk of accidents worldwide are: mining, agriculture, including forestry and logging, and construction.

Chapter 5

FRAMEWORK FOR REGULATING HEALTH AND SAFETY

Throughout the world there are many legal frameworks in place for regulating health and safety, however, if one studies them closely they tend to fall into three broad legal models:

1. **International Labour Organisation (ILO)** – this tends to have a number of common elements to the two others that follow
2. **The American Model** which is prescriptive, and,
3. **The European Model** – this tends to be based around goal setting

Below is more detail of the above three models, we need to understand these principles of the various model's The reason for this is that this thesis is based on 'International' rather than a country specific legislation.

XIII- The ILO Model

Prior to looking at this model the reader needs to understand why the ILO was founded, how it is organised and how it establishes means of compliance.

This is specialised agency of the United Nations system dedicated to improving labour conditions and living standards throughout the world.

Established in 1919 through the Treaty of Versailles as an agency of the League of Nations, the ILO became the first specialised agency affiliated with the UN in 1946.

Its activities include compiling labour statistics, protecting international migrants, and safeguarding trade-union rights.

Its delegates come from member governments (50%), labour (25%), and employers (25%). About 175 countries are represented in the ILO. It received the Nobel Prize for Peace in 1969.

As stated by its Director-General, "the primary goal of the ILO today is to promote opportunities for women and men to obtain decent and productive work, in conditions of freedom, equity, security and human dignity." In working towards this goal, the organisation seeks to promote employment creation, strengthen fundamental principles and rights at work - workers' rights, improve social protection, and promote social dialogue as well as provide relevant information, training and technical assistance.

The ILO hosts the International Labour Conference in Geneva every year in June. At the Conference,

Conventions and Recommendations are crafted and adopted by majority decision. The Conference also makes decisions on the ILO's general policy, work programme and budget.

Each member state is represented at the International Labour Conference by four delegates: two government delegates, an employer delegate and a worker delegate. All delegates have individual voting rights, and all votes are equal, regardless of the population of the delegate's member state. The employer and worker delegates are normally chosen in agreement with the most representative national organisations of employers and workers. Usually, the workers' delegates coordinate their voting, as do the employers' delegates.

Adoption of Conventions

One of the principal functions of the ILO involves setting international labour standards through the adoption of Conventions and Recommendations covering a broad spectrum of labour-related subjects and which, together, are sometimes referred to as the International Labour Code.

Adoption of a Convention by the International Labour Conference allows governments to ratify it, and the Convention then becomes a treaty in international law when a specified number of governments have ratified it.

Ratification of Conventions

The coming into force of a Convention results in a legal obligation to apply its provisions by the nations that have ratified it. Ratification of a Convention is voluntary. Conventions that have not been ratified by member states have the same legal force as Recommendations. Governments are required to submit reports detailing their compliance with the obligations of the Conventions they have ratified.

Every year the International Labour Conference's Committee on the Application of Standards examines a number of alleged breaches of international labour standards. In recent years, one of the member states that has received the most attention is Myanmar / Burma, as the country has repeatedly been criticized for its failure to protect its citizens against forced labour exacted by the army.

Subjects of Conventions

All adopted ILO Conventions are considered international labour standards regardless of how many national governments have ratified them. The topics covered by them cover a wide range of issues, from freedom of association to health and safety at work, working conditions in the maritime sector, night work, discrimination, child labour and forced labour.

1998 Declaration of Fundamental Principles and Rights at Work

In 1998 the International Labour Conference adopted the Declaration on Fundamental Principles and Rights at Work. This Declaration identified four issue areas as "core" or fundamental international labour standards, meaning that any ILO member state should have ratified at least the eight key Conventions, which concern freedom of association, the right to organize and collective bargaining; discrimination; forced labour; and child labour. These core or fundamental standards have all been ratified by the overwhelming majority of ILO member states.

Recommendations

Recommendations do not have the binding force of Conventions, and are not subject to ratification by member countries. Recommendations may be adopted at the same time as Conventions to supplement the latter with additional or more detailed provisions. The intent of these recommendations is often to more precisely detail the principles of related Conventions.

XIV- The American Model

OSHA (Occupational Safety and Health Administration) is a group set up in 1970 to ensure a safe working environment for employees. OSHA is governmentally funded under the U.S. Department of Labour. OSHA's purpose is to create and enforce safety and health standards in the workplace.

OSHA provides many forms of training and education in proper safety and health procedures that it has created. OSHA regulates what is and is not a potential hazard to employees and conducts inspections to ensure that workplaces are OSHA compliant.

OSHA regulations and prescriptive standards save thousands of lives a year and helps to keep all workplaces a healthy and safe place to work.

Examples of the type of prescriptive standard can be found in the OSHA Process Safety Management Standard – 29 CFR 1910.119 the detailed title and summary contents follow:

• Part Number:	1910
• Part Title:	Occupational Safety and Health Standards
• Subpart:	H
• Subpart Title:	Hazardous Materials
• Standard Number:	1910.119 App C
• Title:	Compliance Guidelines and Recommendations for Process Safety Management (Non-mandatory).

Appendix C to §1910.119 -- Compliance Guidelines and Recommendations for Process Safety Management (Non-mandatory)

This appendix serves as a non-mandatory guideline to assist employers and employees in complying with the requirements of this section, as well as provides other helpful recommendations and information. Examples presented in this appendix are not the only means of achieving the performance goals in the standard.

1. ***Introduction to Process Safety Management.***

   The major objective of process safety management of highly hazardous chemicals is to prevent unwanted releases of hazardous chemicals especially into locations which could expose employees and others to serious hazards. An effective process safety management program requires a systematic approach to evaluating the whole process. Using this approach the process design, process technology, operational and maintenance activities and procedures, non-routine activities and procedures, emergency preparedness plans and procedures, training programs, and other elements which impact the process are all considered in the evaluation. The various lines of defence that have been incorporated into the design and operation of the process to prevent or mitigate the release of hazardous chemicals need to be evaluated and strengthened to assure their effectiveness at each level. Process safety management is the proactive identification, evaluation and mitigation or prevention of chemical releases that could occur as a result of failures in process, procedures or equipment.

The process safety management standard targets highly hazardous chemicals that have the potential to cause a catastrophic incident. This standard as a whole is to aid employers in their efforts to prevent or mitigate episodic chemical releases that could lead to a catastrophe in the workplace and possibly to the surrounding community. To control these types of hazards, employers need to develop the necessary expertise, experiences, judgment and proactive initiative within their workforce to properly implement and maintain an effective process safety management program as envisioned in the OSHA standard. This OSHA standard is required by the Clean Air Act Amendments as is the Environmental Protection Agency's Risk Management Plan. Employers, who merge the two sets of requirements into their process safety management program, will better assure full compliance with each as well as enhancing their relationship with the local community.

While OSHA believes process safety management will have a positive effect on the safety of employees in workplaces and also offers other potential benefits to employers (increased productivity), smaller businesses which may have limited resources available to them at this time, might consider alternative avenues of decreasing the risks associated with highly hazardous chemicals at their workplaces. One method which might be considered is the reduction in the inventory of the highly hazardous chemical. This reduction in inventory will result in a reduction of the risk or potential for a catastrophic incident. Also, employers including small employers may be able to establish more efficient inventory control by reducing the quantities of highly hazardous chemicals on site below the established threshold quantities. This reduction can be accomplished by ordering smaller shipments and maintaining the minimum inventory necessary for efficient and safe operation. When reduced inventory is not feasible, then the employer might consider dispersing inventory to several locations on site. Dispersing storage into locations where a release in one location will not cause a release in another location is a practical method to also reduce the risk or potential for catastrophic incidents.

2. *Employee Involvement in Process Safety Management.*

Section 304 of the Clean Air Act Amendments states that employers are to consult with their employees and their representatives regarding the employer's efforts in the development and implementation of the process safety management program elements and hazard assessments.

Section 304 also requires employers to train and educate their employees and to inform affected employees of the findings from incident investigations required by the process safety management program. Many employers, under their safety and health programs, have already established means and methods to keep employees and their representatives informed about relevant safety and health issues and employers may be able to adapt these practices and procedures to meet their obligations under this standard.

Employers who have not implemented an occupational safety and health program may wish to form a safety and health committee of employees and management representatives to help the employer meet the obligations specified by this standard. These committees can become a significant ally in helping the employer to implement and maintain an effective process safety management program for all employees.

3. ***Process Safety Information.***

Complete and accurate written information concerning process chemicals, process technology, and process equipment is essential to an effective process safety management program and to a process hazards analysis.

The compiled information will be a necessary resource to a variety of users including the team that will perform the process hazards analysis as required under paragraph (e); those developing the training programs and the operating procedures; contractors whose employees will be working with the process; those conducting the pre-start up reviews; local emergency preparedness planners; and insurance and enforcement officials.

The information to be compiled about the chemicals, including process intermediates, needs to be comprehensive enough for an accurate assessment of the fire and explosion characteristics, reactivity hazards, the safety and health hazards to workers, and the corrosion and erosion effects on the process equipment and monitoring tools. Current material safety data sheet (MSDS) information can be used to help meet this requirement which must be supplemented with process chemistry information including runaway reaction and over pressure hazards if applicable.

Process technology information will be a part of the process safety information package and it is expected that it will include diagrams of the type shown in Appendix B of this section as well as employer established criteria for maximum inventory levels for process chemicals; limits beyond which would be considered upset conditions; and a qualitative estimate of the consequences or results of deviation that could occur if operating beyond the established process limits. Employers are encouraged to use diagrams which will help users understand the process.

The information pertaining to process equipment design must be documented. In other words, what were the codes and standards relied on to establish good engineering practice. These codes and standards are published by such organisations as the American Society of Mechanical Engineers, American Petroleum Institute, American National Standards Institute, National Fire Protection Association, American Society for Testing and Materials, National Board of Boiler and Pressure Vessel Inspectors, National Association of Corrosion Engineers, American Society of Exchange Manufacturers Association, and model building code groups.

In addition, various engineering societies issue technical reports which impact process design. For example, the American Institute of Chemical Engineers has published technical reports on topics such as two phase flow for venting devices.

This type of technically recognized report would constitute good engineering practice.

For existing equipment designed and constructed many years ago in accordance with the codes and standards available at that time and no longer in general use today, the employer must document which codes and standards were used and that the design and construction along with the testing, inspection and operation are still suitable for the intended use. Where the process technology requires a design which departs from the applicable codes and standards, the employer must document that the design and construction is suitable for the intended purpose.

4. Process Hazard Analysis.

A process hazard analysis (PHA), sometimes called a process hazard evaluation, is one of the most important elements of the process safety management program. A PHA is an organized and systematic effort to identify and analyze the significance of potential hazards associated with the processing or handling of highly hazardous chemicals.

A PHA provides information which will assist employers and employees in making decisions for improving safety and reducing the consequences of unwanted or unplanned releases of hazardous chemicals. A PHA is directed toward analyzing potential causes and consequences of fires, explosions, releases of toxic or flammable chemicals and major spills of hazardous chemicals. The PHA focuses on equipment, instrumentation, utilities, human actions (routine and nonroutine), and external factors that might impact the process. These considerations assist in determining the hazards and potential failure points or failure modes in a process.

The selection of a PHA methodology or technique will be influenced by many factors including the amount of existing knowledge about the process. Is it a process that has been operated for a long period of time with little or no innovation and extensive experience has been generated with its use? Or, is it a new process or one which has been changed frequently by the inclusion of innovative features? Also, the size and complexity of the process will influence the decision as to the appropriate PHA methodology to use. All PHA methodologies are subject to certain limitations. For example, the checklist methodology works well when the process is very stable and no changes are made, but it is not as effective when the process has undergone extensive change. The checklist may miss the most recent changes and consequently the changes would not be evaluated. Another limitation to be considered concerns the assumptions made by the team or analyst. The PHA is dependent on good judgment and the assumptions made during the study need to be documented and understood by the team and reviewer and kept for a future PHA.

The ideal team will have an intimate knowledge of the standards, codes, specifications and regulations applicable to the process being studied. The selected team members need to be compatible and the team leader needs to be able to manage the team and the PHA study. The team needs to be able to work together while benefiting from the expertise of others on the team or outside the team, to resolve issues, and to forge a consensus on the findings of the study and the recommendations.

The application of a PHA to a process may involve the use of different methodologies for various parts of the process. For example, a process involving a series of unit operations of varying sizes, complexities, and ages may use different methodologies and team members for each operation. Then the conclusions can be integrated into one final study and evaluation.

A more specific example is the use of a checklist PHA for a standard boiler or heat exchanger and the use of a Hazard and Operability PHA for the overall process. Also, for batch type processes like custom batch operations, a generic PHA of a representative batch may be used where there are only small changes of monomer or other ingredient ratios and the chemistry is documented for the full range and ratio of batch ingredients.

Another process that might consider using a generic type of PHA is a gas plant. Often these plants are simply moved from site to site and therefore, a generic PHA may be used for these movable plants. Also, when an employer has several similar size gas plants and no sour gas is being processed at the site, then a generic PHA is feasible as long as the variations of the individual sites are accounted for in the PHA.

Finally, when an employer has a large continuous process which has several control rooms for different portions of the process such as for a distillation tower and a blending operation, the employer may wish to do each segment separately and then integrate the final results.

Additionally, small businesses which are covered by this rule, will often have processes that have less storage volume, less capacity, and less complicated than processes at a large facility. Therefore, OSHA would anticipate that the less complex methodologies would be used to meet the process hazard analysis criteria in the standard. These process hazard analyses can be done in less time and with a few people being involved. A less complex process generally means that less data, P&IDs, and process information is needed to perform a process hazard analysis.

Many small businesses have processes that are not unique, such as cold storage lockers or water treatment facilities. Where employer associations have a number of members with such facilities, a generic PHA, evolved from a checklist or what-if questions, could be developed and used by each employer effectively to reflect his/her particular process; this would simplify compliance for them.

When the employer has a number of processes which require a PHA, the employer must set up a priority system of which PHAs to conduct first. A preliminary or gross hazard analysis may be useful in prioritizing the processes that the employer has determined are subject to coverage by the process safety management standard. Consideration should first be given to those processes with the potential of adversely affecting the largest number of employees.

This prioritising should consider the potential severity of a chemical release, the number of potentially affected employees, the operating history of the process such as the frequency of chemical releases, the age of the process and any other relevant factors. These factors would suggest a ranking order and would suggest either using a weighing factor system or a systematic ranking method.

The use of a preliminary hazard analysis would assist an employer in determining which process should be of the highest priority and thereby the employer would obtain the greatest improvement in safety at the facility.

Detailed guidance on the content and application of process hazard analysis methodologies is available from the American Institute of Chemical Engineers' Centre for Chemical Process Safety.

5. ***Operating Procedures and Practices.***

Operating procedures describe tasks to be performed, data to be recorded, operating conditions to be maintained, samples to be collected, and safety and health precautions to be taken. The procedures need to be technically accurate, understandable to employees, and revised periodically to ensure that they reflect current operations.

The process safety information package is to be used as a resource to better assure that the operating procedures and practices are consistent with the known hazards of the chemicals in the process and that the operating parameters are accurate.

Operating procedures should be reviewed by engineering staff and operating personnel to ensure that they are accurate and provide practical instructions on how to actually carry out job duties safely.

Operating procedures will include specific instructions or details on what steps are to be taken or followed in carrying out the stated procedures. These operating instructions for each procedure should include the applicable safety precautions and should contain appropriate information on safety implications. For example, the operating procedures addressing operating parameters will contain operating instructions about pressure limits, temperature ranges, flow rates, what to do when an upset condition occurs, what alarms and instruments are pertinent if an upset condition occurs, and other subjects.

Another example of using operating instructions to properly implement operating procedures is in starting up or shutting down the process. In these cases, different parameters will be required from those of normal operation. These operating instructions need to clearly indicate the distinctions between start-up and normal operations such as the appropriate allowances for heating up a unit to reach the normal operating parameters. Also the operating instructions need to describe the proper method for increasing the temperature of the unit until the normal operating temperature parameters are achieved.

Operating procedures and instructions are important for training operating personnel. The operating procedures are often viewed as the standard operating practices (SOPs) for operations. Control room personnel and operating staff, in general, need to have a full understanding of operating procedures.

If workers are not fluent in English then procedures and instructions need to be prepared in a second language understood by the workers. In addition, operating procedures need to be changed when there is a change in the process as a result of the management of change procedures.

All management of change actions must be coordinated and integrated with current operating procedures and operating personnel must be oriented to the changes in procedures before the change is made. When the process is shutdown in order to make a change, then the operating procedures must be updated before start-up of the process.

Training in how to handle upset conditions must be accomplished as well as what operating personnel are to do in emergencies such as when a pump seal fails or a pipeline ruptures. Communication between operating personnel and workers performing work within the process area, such as non-routine tasks, also must be maintained.

The hazards of the tasks are to be conveyed to operating personnel in accordance with established procedures and to those performing the actual tasks. When the work is completed, operating personnel should be informed to provide closure on the job.

6. ***Employee Training.***

All employees, including maintenance and contractor employees, involved with highly hazardous chemicals need to fully understand the safety and health hazards of the chemicals and processes they work with for the protection of themselves, their fellow employees and the citizens of nearby communities.

Training conducted in compliance with 1910.1200, the Hazard Communication standard, will help employees to be more knowledgeable about the chemicals they work with as well as familiarize them with reading and understanding MSDS. However, additional training in subjects such as operating procedures and safety work practices, emergency evacuation and response, safety procedures, routine and non-routine work authorisation activities, and other areas pertinent to process safety and health will need to be covered by an employer's training program.

In establishing their training programs, employers must clearly define the employees to be trained and what subjects are to be covered in their training. Employers in setting up their training program will need to clearly establish the goals and objectives they wish to achieve with the training that they provide to their employees.

The learning goals or objectives should be written in clear measurable terms before the training begins. These goals and objectives need to be tailored to each of the specific training modules or segments. Employers should describe the important actions and conditions under which the employee will demonstrate competence or knowledge as well as what is acceptable performance.

Hands-on-training where employees are able to use their senses beyond listening, will enhance learning. For example, operating personnel, who will work in a control room or at control panels, would benefit by being trained at a simulated control panel or panels.
This realistic type of training can be very effective in teaching employees correct procedures while allowing them to also see the consequences of what might happens if they do not follow established operating procedures. Other training techniques using videos or on-the-job training can also be very effective for teaching other job tasks, duties, or other important

information. An effective training program will allow the employee to fully participate in the training process and to practice their skill or knowledge. Employers need to periodically evaluate their training programs to see if the necessary skills, knowledge, and routines are being properly understood and implemented by their trained employees. The means or methods for evaluating the training should be developed along with the training program goals and objectives.

Training program evaluation will help employers to determine the amount of training their employees understood, and whether the desired results were obtained. If, after the evaluation, it appears that the trained employees are not at the level of knowledge and skill that was expected, the employer will need to revise the training program, provide retraining, or provide more frequent refresher training sessions until the deficiency is resolved.

Those who conducted the training and those who received the training should also be consulted as to how best to improve the training process. If there is a language barrier, the language known to the trainees should be used to reinforce the training messages and information.

Careful consideration must be given to assure that employees including maintenance and contract employees receive current and updated training. For example, if changes are made to a process, impacted employees must be trained in the changes and understand the effects of the changes on their job tasks (e.g., any new operating procedures pertinent to their tasks). Additionally, as already discussed the evaluation of the employee's absorption of training will certainly influence the need for training.

7. ***Contractors.***

Employers who use contractors to perform work in and around processes that involve highly hazardous chemicals, will need to establish a screening process so that they hire and use contractors who accomplish the desired job tasks without compromising the safety and health of employees at a facility.

For contractors, whose safety performance on the job is not known to the hiring employer, the employer will need to obtain information on injury and illness rates and experience and should obtain contractor references. Additionally, the employer must assure that the contractor has the appropriate job skills, knowledge and certifications (such as for pressure vessel welders). Contractor work method's and experiences should be evaluated. For example, does the contractor conducting demolition work swing loads over operating processes or does the contractor avoid such hazards?

Maintaining a site injury and illness log for contractors is another method employers must use to track and maintain current knowledge of work activities involving contract employees working on or adjacent to covered processes. Injury and illness logs of both the employer's employees and contract employees allow an employer to have full knowledge of process injury and illness experience. This log will also contain information which will be of use to those auditing process safety management compliance and those involved in incident investigations.

Contract employees must perform their work safely. Considering that contractors often perform very specialized and potentially hazardous tasks such as confined space entry activities and non-routine repair activities it is quite important that their activities be controlled while they are working on or near a covered process. A permit system or work authorization system for these activities would also be helpful to all affected employers.

The use of a work authorization system keeps an employer informed of contract employee activities, and as a benefit the employer will have better coordination and more management control over the work being performed in the process area. A well run and well maintained process where employee safety is fully recognized will benefit all of those who work in the facility whether they be contract employees or employees of the owner.

8. ***Pre Start-up Safety.***

For new processes, the employer will find a PHA helpful in improving the design and construction of the process from a reliability and quality point of view. The safe operation of the new process will be enhanced by making use of the PHA recommendations before final installations are completed. P&IDs are to be completed along with having the operating procedures in place and the operating staff trained to run the process before start-up.

The initial start-up procedures and normal operating procedures need to be fully evaluated as part of the pre-start-up review to assure a safe transfer into the normal operating mode for meeting the process parameters.

For existing processes that have been shut down for turnaround, or modification, etc., the employer must assure that any changes other than "replacement in kind" made to the process during shutdown go through the management of change procedures. P&IDs will need to be updated as necessary, as well as operating procedures and instructions.

If the changes made to the process during shutdown are significant and impact the training program, then operating personnel as well as employees engaged in routine and non-routine work in the process area may need some refresher or additional training in light of the changes. Any incident investigation recommendations, compliance audits or PHA recommendations need to be reviewed as well to see what impacts they may have on the process before beginning the start-up.

9. ***Mechanical Integrity.***

Employers will need to review their maintenance programs and schedules to see if there are areas where "breakdown" maintenance is used rather than an on-going mechanical integrity program. Equipment used to process, store, or handle highly hazardous chemicals needs to be designed, constructed, installed and maintained to minimize the risk of releases of such chemicals.

This requires that a mechanical integrity program be in place to assure the continued integrity of process equipment. Elements of a mechanical integrity program include the identification and categorization of equipment and instrumentation, inspections and tests, testing and inspection frequencies, development of maintenance procedures, training of maintenance personnel, the establishment of criteria for acceptable

test results, documentation of test and inspection results, and documentation of manufacturer recommendations as to meantime to failure for equipment and instrumentation.
The first line of defence an employer has available is to operate and maintain the process as designed, and to keep the chemicals contained. This line of defence is backed up by the next line of defence which is the controlled release of chemicals through venting to scrubbers or flares, or to surge or overflow tanks which are designed to receive such chemicals, etc.

These lines of defence are the primary lines of defence or means to prevent unwanted releases. The secondary lines of defence would include fixed fire protection systems like sprinklers, water spray, or deluge systems, monitor guns, etc., dikes, designed drainage systems, and other systems which would control or mitigate hazardous chemicals once an unwanted release occurs. These primary and secondary lines of defence are what the mechanical integrity program needs to protect and strengthen these primary and secondary lines of defences where appropriate.

The first step of an effective mechanical integrity program is to compile and categorize a list of process equipment and instrumentation for inclusion in the program. This list would include pressure vessels, storage tanks, process piping, relief and vent systems, fire protection system components, emergency shutdown systems and alarms and interlocks and pumps. For the categorization of instrumentation and the listed equipment the employer would prioritize which pieces of equipment require closer scrutiny than others.

Meantime to failure of various instrumentation and equipment parts would be known from the manufacturers data or the employer's experience with the parts, which would then influence the inspection and testing frequency and associated procedures. Also, applicable codes and standards such as the National Board Inspection Code, or those from the American Society for Testing and Material, American Petroleum Institute, National Fire Protection Association, American National Standards Institute, American Society of Mechanical Engineers, and other groups, provide information to help establish an effective testing and inspection frequency, as well as appropriate methodologies.

The applicable codes and standards provide criteria for external inspections for such items as foundation and supports, anchor bolts, concrete or steel supports, guy wires, nozzles and sprinklers, pipe hangers, grounding connections, protective coatings and insulation, and external metal surfaces of piping and vessels, etc.

These codes and standards also provide information on methodologies for internal inspection, and a frequency formula based on the corrosion rate of the materials of construction. Also, erosion both internal and external needs to be considered along with corrosion effects for piping and valves.

Some of these inspections may be performed by state or local government inspectors under state and local statutes. However, each employer needs to develop procedures to ensure that tests and inspections are conducted properly and that consistency is maintained even where different employees may be involved. Appropriate training is to be provided to maintenance personnel to ensure that they understand the preventive maintenance program procedures, safe practices, and the proper use and application of special equipment or unique tools that may be required. This training is part of the overall training program called for in the standard.

A quality assurance system is needed to help ensure that the proper materials of construction are used, that fabrication and inspection procedures are proper, and that installation procedures recognize field installation concerns. The quality assurance program is an essential part of the mechanical integrity program and will help to maintain the primary and secondary lines of defence that have been designed into the process to prevent unwanted chemical releases or those which control or mitigate a release. "As built" drawings, together with certifications of coded vessels and other equipment, and materials of construction need to be verified and retained in the quality assurance documentation.

Equipment installation jobs need to be properly inspected in the field for use of proper materials and procedures and to assure that qualified craftsmen are used to do the job. The use of appropriate gaskets, packing, bolts, valves, lubricants and welding rods need to be verified in the field. Also, procedures for installation of safety devices need to be verified, such as the torque on the bolts on ruptured disc installations, uniform torque on flange bolts, proper installation of pump seals, etc. If the quality of parts is a problem, it may be appropriate to conduct audits of the equipment supplier's facilities to better assure proper purchases of required equipment which is suitable for its intended service. Any changes in equipment that may become necessary will need to go through the management of change procedures.

10. Non-routine Work Authorizations.

Non-routine work which is conducted in process areas needs to be controlled by the employer in a consistent manner. The hazards identified involving the work that is to be accomplished must be communicated to those doing the work, but also to those operating personnel whose work could affect the safety of the process. A work authorization notice or permit must have a procedure that describes the steps the maintenance supervisor, contractor representative or other person needs to follow to obtain the necessary clearance to get the job started. The work authorisation procedures need to reference and coordinate, as applicable, lockout / tagout procedures, line breaking procedures, confined space entry procedures and hot work authorisations. This procedure also needs to provide clear steps to follow once the job is completed in order to provide closure for those that need to know the job is now completed and equipment can be returned to normal.

11. ***Managing Change.***

To properly manage changes to process chemicals, technology, equipment and facilities, one must define what is meant by change. In this process safety management standard, change includes all modifications to equipment, procedures, raw materials and processing conditions other than "replacement in kind." These changes need to be properly managed by identifying and reviewing them prior to implementation of the change. For example, the operating procedures contain the operating parameters (pressure limits, temperature ranges, flow rates, etc.) and the importance of operating within these limits.

While the operator must have the flexibility to maintain safe operation within the established parameters, any operation outside of these parameters requires review and approval by a written management of change procedure.

Management of change covers such as changes in process technology and changes to equipment and instrumentation. Changes in process technology can result from changes in production rates, raw materials, experimentation, equipment unavailability, new equipment, new product development, change in catalyst and changes in operating conditions to improve yield or quality.

Equipment changes include among others change in materials of construction, equipment specifications, piping pre-arrangements, experimental equipment, computer program revisions and changes in alarms and interlocks. Employers need to establish means and methods to detect both technical changes and mechanical changes. Temporary changes have caused a number of catastrophes over the years, and employers need to establish ways to detect temporary changes as well as those that are permanent.

It is important that a time limit for temporary changes be established and monitored since, without control, these changes may tend to become permanent.

Temporary changes are subject to the management of change provisions. In addition, the management of change procedures are used to insure that the equipment and procedures are returned to their original or designed conditions at the end of the temporary change. Proper documentation and review of these changes is invaluable in assuring that the safety and health considerations are being incorporated into the operating procedures and the process.

Employers may wish to develop a form or clearance sheet to facilitate the processing of changes through the management of change procedures. A typical change form may include a description and the purpose of the change, the technical basis for the change, safety and health considerations, documentation of changes for the operating procedures, maintenance procedures, inspection and testing, P&IDs, electrical classification, training and communications, pre start-up inspection, duration if a temporary change, approvals and authorization.

Where the impact of the change is minor and well understood, a check list reviewed by an authorized person with proper communication to others who are affected may be sufficient. However, for a more complex or significant design change, a hazard evaluation procedure with approvals by operations, maintenance, and safety departments may be appropriate. Changes in documents such as P&IDs, raw materials, operating procedures, mechanical integrity programs, electrical classifications, etc., need to be noted so that these revisions can be made permanent when the drawings and procedure manuals are updated.

Copies of process changes need to be kept in an accessible location to ensure that design changes are available to operating personnel as well as to PHA team members when a PHA is being done or one is being updated.

12. ***Investigation of Incidents.***

Incident investigation is the process of identifying the underlying causes of incidents and implementing steps to prevent similar events from occurring. The intent of an incident investigation is for employers to learn from past experiences and thus avoid repeating past mistakes. The incidents for which OSHA expects employers to become aware and to investigate are the types of events which result in or could reasonably have resulted in a catastrophic release. Some of the events are sometimes referred to as "near misses," meaning that a serious consequence did not occur, but could have.

Employers need to develop in-house capability to investigate incidents that occur in their facilities. A team needs to be assembled by the employer and trained in the techniques of investigation including how to conduct interviews of witnesses, needed documentation and report writing. A multi-disciplinary team is better able to gather the facts of the event and to analyze them and develop plausible scenarios as to what happened, and why.

Team members should be selected on the basis of their training, knowledge and ability to contribute to a team effort to fully investigate the incident. Employees in the process area where the incident occurred should be consulted, interviewed or made a member of the team. Their knowledge of the events form a significant set of facts about the incident which occurred.

The report, its findings and recommendations are to be shared with those who can benefit from the information. The cooperation of employees is essential to an effective incident investigation. The focus of the investigation should be to obtain facts, and not to place blame. The team and the investigation process should clearly deal with all involved individuals in a fair, open and consistent manner.

13. Emergency Preparedness.

Each employer must address what actions employees are to take when there is an unwanted release of highly hazardous chemicals. Emergency preparedness or the employer's tertiary (third) lines of defence are those that will be relied on along with the secondary lines of defence when the primary lines of defence which are used to prevent an unwanted release fail to stop the release.

Employers will need to decide if they want employees to handle and stop small or minor incidental releases. Whether they wish to mobilise the available resources at the plant and have them brought to bear on a more significant release, or whether employers want their employees to evacuate the danger area and promptly escape to a pre-planned safe zone area, and allow the local community emergency response organisations to handle the release, or whether the employer wants to use some combination of these actions.

Employers will need to select how many different emergency preparedness or tertiary lines of defence they plan to have and then develop the necessary plans and procedures, and appropriately train employees in their emergency duties and responsibilities and then implement these lines of defence.

Employers at a minimum must have an emergency action plan which will facilitate the prompt evacuation of employees when an unwanted release of highly hazardous chemical. This means that the employer will have a plan that will be activated by an alarm system to alert employees when to evacuate and, that employees who are physically impaired, will have the necessary support and assistance to get them to the safe zone as well. The intent of these requirements is to alert and move employees to a safe zone quickly. Delaying alarms or confusing alarms are to be avoided. The use of process control centres or similar process buildings in the process area as safe areas is discouraged. Recent catastrophes have shown that a large life loss has occurred in these structures because of where they have been sited and because they are not necessarily designed to withstand over-pressures from shockwaves resulting from explosions in the process area.

Unwanted incidental releases of highly hazardous chemicals in the process area must be addressed by the employer as to what actions employees are to take. If the employer wants employees to evacuate the area, then the emergency action plan will be activated. For outdoor processes where wind direction is important for selecting the safe route to a refuge area, the employer should place a wind direction indicator such as a wind sock or pennant at the highest point that can be seen throughout the process area. Employees can move in the direction of cross wind to upwind to gain safe access to the refuge area by knowing the wind direction. If the employer wants specific employees in the release area to control or stop the minor emergency or incidental release, these actions must be planned for in advance and procedures developed and implemented.

Preplanning for handling incidental releases for minor emergencies in the process area needs to be done, appropriate equipment for the hazards must be provided, and training conducted for those employees who will perform the emergency work before they respond to handle an actual release. The employer's training program, including the Hazard Communication standard training is to address the training needs for employees who are expected to handle incidental or minor releases.

Preplanning for releases that are more serious than incidental releases is another important line of defence to be used by the employer. When a serious release of a highly hazardous chemical occurs, the employer through preplanning will have determined in advance what actions employees are to take.

The evacuation of the immediate release area and other areas as necessary would be accomplished under the emergency action plan. If the employer wishes to use plant personnel such as a fire brigade, spill control team, a hazardous materials team, or use employees to render aid to those in the immediate release area and control or mitigate the incident, these actions are covered by 1910.120, the Hazardous Waste

Operations and Emergency Response (HAZWOPER) standard.

Responders may be working under very hazardous conditions and therefore the objective is to have them competently led by an on-scene incident commander and the commander's staff, properly equipped to do their assigned work safely, and fully trained to carry out their duties safely before they respond to an emergency. Drills, training exercises, or simulations with the local community emergency response planners and responder organisations is one means to obtain better preparedness. This close cooperation and coordination between plant and local community emergency preparedness managers will also aid the employer in complying with the Environmental Protection Agency's Risk Management Plan criteria.

One effective way for medium to large facilities to enhance coordination and communication during emergencies for on plant operations and with local community organisations is for employers to establish and equip an emergency control centre. The emergency control centre would be sited in a safe zone area so that it could be occupied throughout the duration of an emergency.

The centre would serve as the major communication link between the on-scene incident commander and plant or corporate management as well as with the local community officials. The communication equipment in the emergency control centre should include a network to receive and transmit information by telephone, radio or other means.

It is important to have a backup communication network in case of power failure or one communication means fails. The centre should also be equipped with the plant layout and community maps, utility drawings including fire water, emergency lighting, appropriate reference materials such as a government agency notification list, company personnel phone list, SARA Title III reports and material safety data sheets, emergency plans and procedures manual, a listing with the location of emergency response equipment, mutual aid information, and access to meteorological or weather condition data and any dispersion modelling data.

14. Compliance Audits.

Employers need to select a trained individual or assemble a trained team of people to audit the process safety management system and program. A small process or plant may need only one knowledgeable person to conduct an audit. The audit is to include an evaluation of the design and effectiveness of the process safety management system and a field inspection of the safety and health conditions and practices to verify that the employer's systems are effectively implemented.

The audit should be conducted or lead by a person knowledgeable in audit techniques and who is impartial towards the facility or area being audited. The essential elements of an audit program include planning, staffing, conducting the audit, evaluation and corrective action, follow-up and documentation.

Planning in advance is essential to the success of the auditing process. Each employer needs to establish the format, staffing, scheduling and verification methods prior to conducting the audit. The format should be designed to provide the lead auditor with a procedure or checklist which details the requirements of each section of the standard. The names of the audit team members should be listed as part of the format as well. The checklist, if properly designed, could serve as the verification sheet which provides the auditor with the necessary information to expedite the review and assure that no requirements of the standard are omitted. This verification sheet format could also identify those elements that will require evaluation or a response to correct deficiencies. This sheet could also be used for developing the follow-up and documentation requirements.

The selection of effective audit team members is critical to the success of the program. Team members should be chosen for their experience, knowledge, and training and should be familiar with the processes and with auditing techniques, practices and procedures. The size of the team will vary depending on the size and complexity of the process under consideration. For a large, complex, highly instrumented plant, it may be desirable to have team members with expertise in process engineering and design, process chemistry, instrumentation and computer controls, electrical hazards and classifications, safety and health disciplines, maintenance, emergency preparedness, warehousing or shipping, and process safety auditing. The team may use part-time members to provide for the depth of expertise required as well as for what is actually done or followed, compared to what is written.

An effective audit includes a review of the relevant documentation and process safety information, inspection of the physical facilities, and interviews with all levels of plant personnel. Utilizing the audit procedure and checklist developed in the preplanning stage, the audit team can systematically analyze compliance with the provisions of the standard and any other corporate policies that are relevant. For example, the audit team will review all aspects of the training program as part of the overall audit.

The team will review the written training program for adequacy of content, frequency of training, effectiveness of training in terms of its goals and objectives as well as to how it fits into meeting the standard's requirements, documentation, etc. Through interviews, the team can determine the employee's knowledge and awareness of the safety procedures, duties, rules, emergency response assignments, etc. During the inspection, the team can observe actual practices such as safety and health policies, procedures, and work authorization practices. This approach enables the team to identify deficiencies and determine where corrective actions or improvements are necessary.

An audit is a technique used to gather sufficient facts and information, including statistical information, to verify compliance with standards. Auditors should select as part of their preplanning a sample size sufficient to give a degree of confidence that the audit reflects the level of compliance with the standard. The audit team, through this systematic analysis, should document areas which require corrective action as well as those areas where the process safety management system is effective and working in an effective manner. This provides a record of the audit procedures and findings, and serves as a baseline of operation data for future audits. It will assist future auditors in determining changes or trends from previous audits.

Corrective action is one of the most important parts of the audit. It includes not only addressing the identified deficiencies, but also planning, follow-up, and documentation. The corrective action process normally begins with a management review of the audit findings. The purpose of this review is to determine what actions are appropriate, and to establish priorities, timetables, resource allocations and requirements and responsibilities. In some cases, corrective action may involve a simple change in procedure or minor maintenance effort to remedy the concern.

Management of change procedures need to be used, as appropriate, even for what may seem to be a minor change. Many of the deficiencies can be acted on promptly, while some may require engineering studies or in-depth review of actual procedures and practices. There may be instances where no action is necessary and this is a valid response to an audit finding. All actions taken, including an explanation where no action is taken on a finding, will need to be documented as to what was done and why.

It is important to assure that each deficiency identified is addressed, the corrective action to be taken noted, and the audit person or team responsible be properly documented by the employer. To control the corrective action process, the employer should consider the use of a tracking system.

This tracking system might include periodic status reports shared with affected levels of management, specific reports such as completion of an engineering study, and a final implementation report to provide closure for audit findings that have been through management of change, if appropriate, and then shared with affected employees and management.

This type of tracking system provides the employer with the status of the corrective action. It also provides the documentation required to verify that appropriate corrective actions were taken on deficiencies identified in the audit.The above appears to be quite detailed for the reader – but the detail has been included to give a better understanding of the term 'prescriptive'.

XV- The European Model

The European Union is an ever-evolving alliance of more than fifteen European countries designed to foster economic cooperation among its members. With its roots stretching back to just after World War II, this alliance has the ultimate goal of unifying the economic interests of these countries in order to reduce the chance of widespread armed conflict returning to the European continent.

History - In the early 1950s, Germany and France spear headed the establishment of the European Coal and Steel Community. At this time, West Germany was in the process of rebuilding its war ravaged steel industry under the direction of the United States and Great Britain. This was an understandable source of concern for France, as German industrial might had been used against the French in the two world wars earlier in the century. Consequently, a federation that would govern these important economic resources won the approval of both the French and Germans. They were also joined by Italy, Belgium, Luxembourg, and the Netherlands as the original six founding nations of what would eventually become the European Union.

The European Coal and Steel Community was later followed by the establishment of the European Economic Community under the Treaty of Rome, which was signed by the same six countries in 1957. This treaty established the framework for the six member countries to pursue an economic and monetary union by creating a single market to further their economic development. The European Economic Community established a customs union for re moving trade barriers, such as tariffs and quotas, between member countries over a period of several years. Also, common external tariffs were phased in for goods entering the union.

The single market established by the European Economic Community also opened the door for the free movement of workers, businesses, and capital throughout the community. Border stops and customs checks were eliminated, companies expanded across national borders, and financial institutions expanded with them.
Within the European Union (EU), the sovereignty and right of each Member State to pass laws on safety and health, or on any subject, is not affected by the membership.

By agreement the Member States harmonise their regulatory activities by discussing regulatory action in the European forum, resulting in the issue of Directives. These contain the principles and objectives that have to be attained by the national legislation within each of the Member States.

In addition there is also a timetable for the implementation of the said legislation, which has to be sufficient to ensure compliance with the minimum standards set out in the directives. It has to be said that national needs do vary the extent to which internal legislation in the Member States exceed or extend or even vary the Directives.

Examples of relevant European Safety and Health Directives include:

Council Directive 90/269/EEC (EU) of 29 May 1990 on the minimum health and safety requirements for the manual handling of loads where there is a risk particularly of back injury to workers (fourth individual Directive within the meaning of Article 16 (1) of Directive 89/391/EEC)

Council Directive 92/58/EEC of 24 June 1992 on the minimum requirements for the provision of safety and/or health signs at work (ninth individual Directive within the meaning of Article 16 (1) of Directive 89/391/EEC)

Council Directive 1999/92/EC of the European Parliament and of the Council of 16 December 1999 on minimum requirements for improving the safety and health protection of workers potentially at risk from explosive atmospheres (15th individual Directive within the meaning of Article 16(1) of Directive 89/391/EEC

Council Directive 93/104/EC of 23 November 1993 concerning certain aspects of the organisation of working time

Council Directive 89/655/EEC of 30 November 1989 concerning the minimum safety and health requirements for the use of work equipment by workers at work (second individual Directive within the meaning of Article 16 (1) of Directive 89/391/EEC Official Within the notes it is worth mentioning the United Kingdom (UK), as it is a member of the EU. Within the UK a fundamental principle of their health and safety legislation is that the responsibility for compliance lies with those who own, manage and work in industrial and commercial businesses.

The most significant European Union Directive impacting the UK is the Framework Directive (89/391/EEC) for the Introduction of Measures to Encourage Improvements in the Safety and Health of Workers, widely referred to as the 'Framework Directive'. This directive was adopted by the Council of Ministers in 1989 and was required to be implemented into the domestic legislation of Member States by 1 January 1993.

The directive placed new duties upon the employers, including those:

1. To assess health and safety risks.
2. To develop a prevention policy.
3. To designate competent personnel. To make arrangements for emergency procedures.

4. To provide information for workers.

5. To allow workers' representatives time off with pay for appropriate training.

6. To give workers a right to stop work when faced with imminent danger.

The greater part of the Framework Directive (including the key general duty to assess risk) has been implemented by the Management of Health and Safety at Work Regulations 1992 (S.I. 1992, No. 2051). This was replaced by the Management of Health and Safety at Work Regulations 1999 (S.I. 1999, No. 3242) on 29 December 1999.

It is interesting to note that almost all of the health and safety risks arising from work activities within the UK are regulated within a single flexible legal framework, in which the UK Parliament passes health and safety related laws as either

a) Acts – e.g. The Health and Safety at Work etc Act 1974 (HASAWA) and

b) Regulations – e.g. The Management of Health and Safety at Work Regulations 1999 (MHSWR)

HASAWA was a watershed development in health and safety law. It implemented the major recommendations of the Robens Report, i.e. the establishment of a unified structure of administration and enforcement of health and safety legislation and the creation of general goal-setting duties for employers.

The HASAWA was therefore important in a number of respects:

1. It unified the administration and enforcement of health and safety legislation.

2. It set out the principles of health and safety law reduced to a number of general goal setting duties.

3. Being an enabling statute, it created a mechanism for the creation of a raft of supporting regulations. Since HASAWA most new legislation has taken the form of regulations made under that Act.

4. It created a distinct division between the civil and criminal aspects of health and safety law.

Up until the HASAWA breaches of health and safety legislation, generally, gave rise to a criminal offence as well as creating a civil cause of action for the benefit of anyone injured as a result of the breach. The consequences of breaching the 1974 Act, almost exclusively, give rise to enforcement and regulatory issues and are not concerned with civil claims for damages.

Since the creation of the HASAWA back in 1974 there has since then been a gradual progressive reform of the whole legal system which is resulting in a move away from prescriptive industry specific legislation to the use of goal setting legislation expressing such goals and general principles. A good example here is the MHSAWR, which created the legal framework for the completion of risk assessments.

Within the UK there are two non-departmental Government bodies with specific statutory functions related to health and safety, these are:

a) The Health and Safety Commission (HSC) and

b) The Health and Safety Executive (HSE)

The HSC is appointed by the respective Secretary of State. The HSC has the statutory responsibility under the HASAWA for proposing laws relating to health and safety as well as standards to the UK Government.

In the preparation of such proposals the HSC relies on advice from the HSE as well as feedback from various consultation exercises with a broad range of organisations and individuals that represents the interests and views of businesses, professions and workers.

Enforcement

For a whole host of reasons companies as well as individuals do no comply with the requirements of legislation or legislation, as a consequence persons and other aspects are put at risk. In order to ensure legal compliance governments need to consider how laws can be complied with and the legislation / regulation 'enforced' via enforcement authorities who employ officers with respective enforcement powers.

Enforcement authorities usually have a variety of powers that enable them to deal with breaches (non compliance) of legislation and other situations that they consider to be unsafe. These powers can result in enforcement action, the extent of which depends on the gravity of the non-compliance.

Enforcement action can take several paths:

a) Informal – here examples would include the provision of health and safety related advice and guidance in either verbal or written form, and

b) Formal – here one or a combination of the following could be used:

   - Issuing of a formal caution – this may be used as evidence at a latter date – this aspect tends to be used for the more minor repeat offences / breaches
   - Issuing legal mandates to comply with legislation – these can take the form of improvement notices
   - Issuing legal mandates to stop dangerous activities – these can take the form of prohibition notices
   - Prosecution – here the intent would be to punish in some way the wrong doer – company or

individual with court orders to remedy a given situation, and or fines and or ultimately imprisonment. In some countries serious breaches of legislation still have corporal punishment available as an ultimate option.

UK Examples

Within the UK today there is a growing trend towards more prosecutions of individuals, particularly senior management (including directors) where due to their connivance, consent or negligence an offence has been committed whereby legal compliance has not been achieved.

Above the topic of improvement and prohibition notices was briefly mentioned. At this juncture it would be useful for the reader to have a little more detail on these topics.

Improvement Notice

An Improvement Notice may be served where an inspector is of the opinion that a person:

1. Is contravening one or more statutory provisions, or
2. Has done so in the past, and it is likely that the contravention will continue or be repeated.

An Improvement Notice will give reasons as to why an inspector has formed this view, and will require the person to remedy the alleged contravention within a specified time period, which will be at least 21 days following service of the notice.

An appeal can be made on one of three grounds, against:

1. The service of the notice
2. The contents of the notice
3. The time limit given for compliance.

An industrial tribunal can then do one of three things:

1. Cancel the notice
2. Affirm it
3. Affirm it with modifications.

Bringing an appeal suspends an Improvement Notice until either the appeal is heard, or withdrawn.

Prohibition Notices

A Prohibition Notice can be served by an inspector if he is of the opinion that work activities being carried out under the control of a person will involve a risk of serious personal injury. There is no need for any specific statutory provisions to have been breached, although in practice this will almost invariably be the case. The existence of the risk of serious personal injury is in itself sufficient to justify service of a Prohibition Notice.

The effect of the notice is to prevent the activity which is specified in it from being carried on, until matters are remedied to the inspector's satisfaction.

A Prohibition Notice should specify what matters the inspector considers give rise to a risk of serious personal injury. If the activities do involve a breach of legislation, the inspector should specify what that is, and why he is of the opinion that the legislation has been breached. The activities specified in the notice cannot be resumed until the inspector is satisfied that the risk of personal injury has been removed.

A Prohibition Notice will usually take immediate effect, but can be deferred to take effect at the end of a period specified within the notice i.e. in circumstances of certain chemical processes where an immediate stop would not be safe.

An appeal can be made against one of three things:

1. The service of a Prohibition Notice
2. The contents
3. The time limit given for compliance.

An industrial tribunal can do one of three things:

1. Cancel the notice
2. Affirm it
3. Affirm it with modifications.

Bringing an appeal does not suspend the Prohibition Notice. If the person served with the notice wishes to have it suspended, an application must be made to the industrial tribunal in advance of the appeal being heard.

It is an offence to contravene any requirement or prohibition imposed by an Improvement or Prohibition Notice (including where the notice has been modified on appeal).

Implementation of Legislation

Throughout the world there are differences to be seen with the methods in which legislation is implemented with regard to the standards of enforcement. The many reasons behind this verification between different countries are wide and diverse and may have several causes, including one or more of the following:

- Differing legal systems – as discussed earlier in prescriptive or goal setting
- The differing standards of legislation enforcement
- The penalties available for legislative breaches
- Cultural or religious beliefs and how different societies view the potential offence(s)
- The knowledge and competence of the Enforcing Authorities and those working for them

- The workload of the Enforcement Officers and the funding that the function receives
- The effectiveness and degree of monitoring undertaken by the Enforcement Authorities
- The methods and effectiveness of reporting likely offices to Enforcing Authorities
- Enforcement initiatives – political pressure to enforce – international pressure

Consistency in approach to the above areas is seen as an important goal for any responsible company, government, or alliance of governments.

The variation in enforcement within any identified individual country or countries may be determined by one or more of the following factors:

1. Enforcement of Standards Through The Legal System

   To be effective any enforcement program needs a number of key elements to be satisfied i.e.

   - An adequate number of trained and competent regulators
   - An adequate number of trained and competent inspectors who's duties extend to the planned and random inspections of the workplaces covered by the respective legislation

- The investigation of complaints and incidents, and
- Conducting formal proceedings as necessary following the previously mentioned inspections and or investigations

For traditional and financial reasons it is not uncommon to find the enforcement of health and safety within the framework of a Department of Labour, here though the Departments Inspectors are also engaged with policies relating to wages, conditions, etc. As a consequence the resources of these Inspectors will be stretched in a bid to accommodate the specialist knowledge required of health and safety related issues.

2. Advice to Employers, Workers and The Public
   Here such advice would be related to the appropriate control measures required to be implemented and the issue of explanatory publications. Such a resource development of information in topics such as the control of hazards, assessment of risk and the collection and collation of accident, injury and ill-health data into an information system which in turn allows a focus to answer problems within a particular community and or to monitor the effectiveness of national measures in force
3. Uniformity of Approach

It goes without saying that for all practical reasons there must be an appropriate and fair means of tackling those persons or organisations who wilfully and deliberately avoid national legislative compliance. Also any penalties imposed for such breaches should be appropriate to the offence and set at a common level for the same offence committed in a different place and or environment.

The level of the penalty applied should be such that there is no economic advantage to be gained by any failure to comply, followed by a willingness to accept any resulting penalty.
Equally there needs to be some relation between national control measures and those of other international trading partners and other countries with broadly similar economies and standards. This is a significant consideration of multi-national employers.

4. Target Setting
Here Governments are relied upon to introduce initiatives to improve the future of national performance. The elements that can be included here rage from

- The requirement of a formal risk assessment as part of a planning process, making designers responsible for the provision of information relating to their designs and the future maintenance of buildings (including dismantling), plant and equipment.

- The publication of injury and ill-health data and other results set against future targets

Governments who do not undertake a progressive assessment of the impact measures on persons lives and well-being of the workforce are seen in the international arena as merely playing lip service to the common international goal in the reduction of the number of workplace accidents (of all kinds).

5. Fostering Co-Operation

Here there is a need to mention the fostering of co-operation between all of the parties involved, this would include workers, their representatives, managers and their representatives and of course government.

Today, a number of countries have a legislative system that provides for joint consultation and a process of information exchange on subjects relating to workplace health and safety standards and the state of workers safety and health. Legislation that has the support of such a system involving these parties is known to be more likely to be effective that that of those which are developed in isolation.

Experience in many countries is that the potential for the contribution to good working and workplace practices by workers is quite often undervalued. It is know and research has shown that those workers who are consulted on their own safety and health are more likely to be co-operative with measures that are put in place to protect them.

International and National Standards

Research clearly shows that the term "occupational safety and health" refers to two distinct types of standards, being:

- Labour Standards – Here the focus is related to the general occupational safety and health conditions in the workplace that are present to protect the worker(s), and

- Product Standards – These tend to focus on the duties of occupational safety and health suppliers of substances, articles, etc and the avoidance of competitive advantage i.e. allowing the lowering of safety standards by one company to be able to give a preferential price over and against another who sells a similar machine complying with legislative requirements.

International Standards

The standards defined in the legislation of any country relating to safety and health should at least meet and preferably exceed the internationally accepted minimum standard for each aspect.

An organisation working within that country should themselves, as a minimum meet and preferably exceed the standards set in the legislation of that country and be adopting this as best practice. It follows that by adopting best practiced standards organisations are now providing improved levels of protection to and for their employees. Such organisations tend to be regarded as responsible and caring employers, this in turn often gives them a competitive advantage in the marketplace, as such morals become an increased focus of the international community.

When researching best practice standards what tends to come to the fore is that these practices tend to be developed by industry groups, professional bodies and at times international organisations.

The ILO Conventions and Recommendations were discussed earlier in this study element and if the reader cares to research the topic further he will find that there is ILO Labour and Social Guidance material available in the form of 'codes of practice', an example being ILO - OSH Guidelines of Occupational Safety and Health Management Systems. There are also associated reference manuals available.

In the ever changing commercial world we see trade tariffs being reduced or eliminated altogether, and there is a greater occurrence of regional economic integration agreements being signed as part of multilateral trade agreements within the framework of the World Trade Organisation (WTO). It is fair to say that non-tariff technical barriers tend to gain more significance in matters of international trade.

When one studies the WTO publication "Technical Barriers to Trade Agreement" it is clear to see the harmonisation of technical product standards. This harmonisation being based on the adherence of the national standards to international standards, and the transparency of the standard setting process. developing countries.

When countries adapt to TBT/1994 the effort exerted will bring about benefits, the two primary ones being:

- Increased opportunities for the export of their conforming products, and
- A significant improvement in the safety of those products.

Once a harmonised standard has been adopted, the International Standards Organisation (ISO) publishes international standards, there are many examples of these, a couple being:

- ISO 12100 – 1 - 2003 – Safety of Machinery, Basic Concepts, General Principles for Design, Basic Terminology and Methodology, and
- ISO 12100 – 2 - 2003 – Safety of Machinery, Basic Concepts, General Principles for Design, Technical Principles.

The European Model

European Harmonised (Europe'en de Normalisation (EN)), technical standards have been written by the EU member state committees consisting of standards bodies e.g. The UK British Standards Institute (BSI).

Within the EU the two most common types of European standards relating to safety and health are:

a. Comit'e Europe'en de Normalisation (CEN) – this considers machinery, and

b. Comit'e Europe'en de Normalisation Electrotechnique (CENELEC) – concentrating on electrical equipment

Once adopted, the EN standard is referred to as a 'transposed harmonised standard'. Here for example EN 394 – Safety of Packaging Machines applies to the whole of the EU and would take precedence over any national standard covering the same subject.

National Standards

Within the UK the EN technical standards are adopted by the BSI, and are published as a BS EN, using the example quoted above EN 394, would be referred to as BS EN 394 – Safety of packaging machines.

With the adoption of international standards the UK standards being published as ISO standards would be published as BS EN ISO 12100, again using earlier references in this training element.

Within the UK practical detail on legislation is provided by primary (key) supporting documents, these are referred to as Approved Codes of Practice (ACoP) and Guidance.

Approved Code of Practice

Legislation allows for the development and approval of statements of preferred work practice, known as "approved codes of practice".

These are recommended means of compliance with the requirements of particular pieces of legislation, and are developed normally after consultation with the industry or industries concerned. They are approved by the respective Government Minister after consultation with affected groups and individuals.

A code of practice applies to anyone who has a duty of care in the circumstances described in the code - which may include employers, employees, the self-employed, principals to contracts, owners of buildings or plant, and so on.

An approved code does not necessarily contain the only acceptable ways of achieving the standard required by legislation. But, in most cases, compliance will meet the requirements of the legislation, in relation to the subject matter of the code.

An approved code does not have the same legal force as a regulation, and failure to comply with a code of practice is not, of itself, an offence. However, observance of a relevant code of practice may be considered as evidence of good practice in a court.

Guidance

Guidance is written by any number of responsible organisations such as the HSC, HSE, Trade Bodies, BSI, etc. If an organisation follows the guidance laid down they would normally be considered as doing enough to comply with legislation.

Guidance has:

- Practical advice on how compliance with legislation can be achieved
- There tends to be more discussion than that within an ACoP

- No legal standing, but as said earlier, full compliance would normally be viewed as being evidence of adhering to good practice in a court of law.

Sources of Information

Whenever viewing documentation relating to the duties of employers it will be seen that a general duty placed on the employer is for them to provide their workers / employees with relevant health and safety information. This information should at least cover the following:

a. The risks of the work in which the person is exposed

b. The risk control measures both preventative and precautionary

c. Emergency procedures and protocols to be adopted i.e. fire, accident, major spillage, etc

d. Risks from third parties i.e. contractors

e. Names of nominated / responsible persons i.e. First Aiders, Fire Wardens

f. Names of Workplace / Worker Safety Representative (where these are in place)

Importance of Information

The provision of information is important for many reasons and the information necessary for a person to undertake their work safely and effectively comes in many forms. There will, for example, be operating instructions for equipment as well as signs and warning labels. It is particularly important that any information with which the worker is provided must be in a form that is easy to understand and presented in such a way that it is actually taken notice of.

There are also aspects of information provision that are more intimately connected with the work at hand, such as reading screens, dials and meters on equipment, which are also important aspects of information provision. However, this topic will only be concerned with the more general aspects of the means of information provision

Key Factors for Information Provision

Most Important Factors

The most important aspects of the provision of information is that it is:

1. Visible - i.e. easily seen and not torn, dirty, hidden or mixed with large amounts of irrelevant information.
2. Clear - i.e. the meaning of the information (or warning) must be immediately understandable to the group of people for whom it is intended, particularly if this includes members of the public.
3. Relevant - i.e. the information should be relevant and directly applicable to the area in which it is sited. For example, a warning notice or sign must be as close as possible to the item, area or equipment to which the warning applies.

Feedback and Clarification

One particular aspect of information provision is the need to ensure that the intended recipient of the information actually correctly understands the intended meaning. In the case of information provision during training courses, briefings or meetings it is possible for the attendees (although not always easy for some) to ask questions to clarify an issue. However, in the case of purely visual information provision such as signs, posters and notices, the opportunity for direct clarification is completely absent.

As a result it is important to ensure that signs are instantly recognisable and notices, posters and other forms of visually provided information include contact details to enable anyone with questions to easily seek feedback.

The inclusion of contact information can also have a secondary benefit, as it will also give an indication of the authority on which the information is given.

Timely Relevance

A particularly important aspect of visual means of information is to ensure that if it the information is available, e.g. a poster is still on display, then the it must still be valid. In all too many instances visual information, such as signs and notices, are not removed when a particular time-dependant situation has changed.
As a minor example, consider how many people ignore 'wet paint' signs because their experience is that the warning signs are left up for days after the paint has dried. It is important to ensure either that out-of-date information is removed or, at the very least, actually has a date or time after which it may be safely ignored. If this rule is strictly adhered to, the signs, notices and other information that is available to workers is much more likely to be believed and complied with.

From Where Can Information Be Obtained?

Advice and information on the control of workplace hazards and risks is required for a safe working environment. Such information can be obtained internally (from within the organisation itself), or externally (outside the organisation).

Internal Information

From within any organisation the potential for available information can be extensive, and these information sources become abundant the more organised an organisation is with regards to its occupational health and safety.

Information, very useful information is available form the following internal sources:

- Policies
- Inspection Reports

- Risk Assessments
- Method Statements
- Accident Records
- Medical Records
- Health and Safety Committee Meeting Minutes
- Health and Safety Officer
- Plant Registers
- In-House Safety Library, etc

External

There is such a lot of health and safety information that is available externally to any organisation. Some of this information can be quite specific in meeting the needs of the persons requiring such information. External information sources for health and safety related matters include:

- Government Bodies i.e. UK HSE, USA OSHA
- International Bodies i.e. ILO, WHO
- Professional Bodies i.e. UK RoSPA, IOSH
- Trade Journals
- Suppliers i.e. for Substances
- Manufacturers i.e. for Equipment
- Insurance Organisations
- Consultants
- Trade Unions / Workers Associations
- The Internet.

While on the subject of the internet a number of home pages to various web sites will be shown:

Health and Safety Management

As stated earlier in this Element some countries do not have a strong and robust system of occupational safety and health law. In such countries the emphasis is on the management of health and safety.
Management by definition is the process of setting policy, organising and controlling the workforce, monitoring and reviewing the efforts and outcomes of an organisation and the use of other organisational resources to achieve stated goals.

With regard to health and safety management these goals will include:

- The protection of the individual (employee or non-employee) from harm, injury or ill-health
- Compliance with legislative requirements to avoid prosecution and to demonstrate best practice
- The cost effective management of health and safety in harmony with other key business objectives

The performance of an organisation, whether looked at in terms of safety alone or more widely in terms of its business measures, is affected by a number of issues. Probably one of the most important aspects is the effectiveness of the management processes within the organisation. However, even the best management processes can be defeated by the way in which people approach and undertake their daily tasks.

A safety management system that has addressed all the risks by creating procedures and safe systems of work is little more than wasted effort if they are not carried out and complied with. Consequently, human factors are the key to the efficient operation of an organisation, and addressing the issues that affect human factors is most likely to gain great return for the effort expended on it.

Many organisations all too often sadly fail to manage their health and safety performance effectively. Many give this the reason as they see it as something distinct or remote from key business functions and other systems of management such as quality management (ISO 9000 series) and Environmental Management (14000 Series)

There are a number of health and safety management systems that exist around the world, these include:

- ILO – OSH, 2001 'Guidelines on Occupational Safety and Health Management' Systems' (ILO)
- OSHAS 18000, 1999 'Occupational Health and Safety Assessment Series'. (BSI) and
- HSG 65, 2003. 'Successful Health and Safety Management Systems'. (UK HSE)

The value of Management Systems

The management attitude and systems is as much a condition the culture of an organisation as the organisational culture exerts an effect on those systems. A poorly planned management system that forces unreasonable demands on a workforce is more likely to promote a negative culture than a positive one.

Where systems are complicated or difficult to carry out - especially where the workforce do not know the reasons for the complexity - they are most likely to fail through lack of compliance.

Similarly, where the resources to carry out the tasks properly are unavailable, there is again a high likelihood that this will cause the reduction of the safety culture. The lack of resources will diminish the workforce's perception of management commitment and will, either through frustration or a misguided attempt to complete the task, cause any positive safety culture in place to dissipate and become increasingly negative and thus render any safety management systems ineffective.

Good management systems, both safety-related and general, are a main contributor to a positive safety culture, and the positive safety culture reinforces the good management systems.

ILO-OSH, 2001 (ILO)

The ILO – OSH, 2001 Document gives a range of requirements for the voluntary application and systematic implementation of occupational safety and health (OSH), management systems. The purpose of designing such a document was as a practical tool that would assist organisations and other competent institutions as a means to achieving continual improvement in OSH performance. The ILO-OSH guidelines are voluntary and as such carry no legal obligation, but they do emphasise a number of concepts such as continual improvement, employers' leadership and commitment, and workers participation.

The ILO-OSH system requires:

a) A competent institution or institutions should be nominated, as appropriate, to formulate, implement and periodically review a coherent national policy for the establishment and promotion of OSH management systems in *organisations.* This should be done in consultation with the most representative organisations of employers and workers, and with other bodies as appropriate.

b) Tailored guidelines, reflecting the overall objectives of the ILO guidelines, should contain the generic elements of the national guidelines and should be designed to reflect the specific conditions and needs of *organisations* or groups of *organisations,* taking into consideration particularly:

   - their size (large, medium and small) and infrastructure; and
   - the types of hazards and degree of risks.

The links between the national framework for OSH management systems (OSH-MS) and its essential elements are illustrated in figure below:

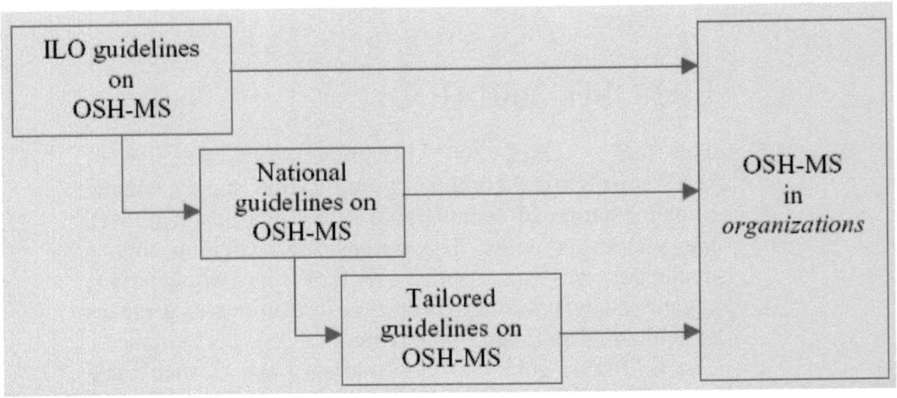

Occupational safety and health, including compliance with the OSH requirements pursuant to national laws and regulations, are the responsibility and duty of the employer.

The OSH Management System should contain the main elements of:

- Policy
- Organisation
- Planning and Implementation
- Evaluation, and
- Action for improvement.

The above is best illustrated with a diagram extracted from the actual ILO-OSH 2001, document, see on the next page.

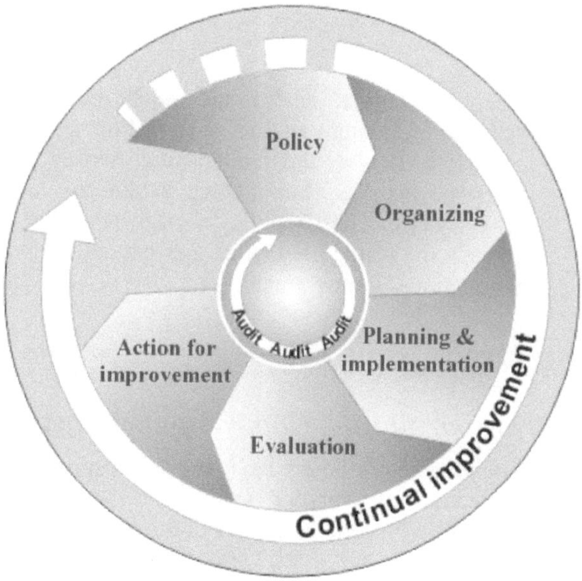

We shall go on to discuss each of the above elements in a little more detail

Policy - Occupational Safety and Health Policy

The employer, in consultation with workers and their representatives, should set out in writing an OSH policy, which should be:

a) specific to the organisation and appropriate to its size and the nature of its activities;

b) concise, clearly written, dated and made effective by the signature or endorsement of the employer or the most senior accountable person in the organisation;

c) communicated and readily accessible to all persons at their place of work;

d) reviewed for continuing suitability; and

e) made available to relevant external interested parties, as appropriate.

The occupational health and safety policy is really a statement of intent of the organisation as their plan for safety. It really is the fundamental starting point for the complete process of managing health and safety. To be properly effective this policy should set a clear direction to achieve the principle goal of loss prevention with the aim for continual improvement.

Organising

Responsibility, Control and Accountability

The employer should have overall responsibility for the protection of workers' safety and health, and provide leadership for OSH activities in the organisation.

A person or persons at the senior management level should be appointed, where appropriate, with responsibility, accountability and authority for:

a) the development, implementation, periodic review and evaluation of the OSH management system;

b) periodic reporting to the senior management on the performance of the OSH management system; and

c) promoting the participation of all members of the organisation.

By committing to the above the organisation starts to establish the correct attitude for a ‘safety culture. Several key considerations need to be embedded to further advance this culture, being:

- Competence and Training – OSH competence includes recruitment or development of the right education level / awareness, ability tied with work experience and training, with qualifications as appropriate or a combination of these for all staff.

- Commitment and Control – – Best demonstrated by the effective allocation of responsibilities and accountabilities that suitably secure commitment from all within the organisation. This being complemented by the provision of training, instruction and supervision to achieve effective control within that organisation.
- Co-Operation – Needs to be ensured between individuals, working groups and departments. This is meant to include consultation on health and safety matters between management and workplace representatives (or workers as appropriate), as well as between the organisation and its contractors, suppliers and clients, etc.
- Communication – This takes all known forms, using the most effective means for any given situation. Such will include oral and written communication and well as visible means and example, probably potentially the strongest communication tool.

Planning and Implementation

The organisation's existing OSH management system and relevant arrangements should be evaluated by an initial review, as appropriate. In the case where no OSH management system exists, or if the organisation is newly established, the initial review should serve as a basis for establishing an OSH management system.
The initial review should be carried out by competent persons, in consultation with workers and / or their representatives, as appropriate. It should:

a) identify the current applicable national laws and regulations, national guidelines, tailored guidelines, voluntary programmes and other requirements to which the organisation subscribes;

b) identify, anticipate and assess hazards and risks to safety and health arising from the existing or proposed work environment and work organisation; and

c) determine whether planned or existing controls are adequate to eliminate hazards or control risks; and

d) analyse the data provided from workers' health surveillance.

The result of the initial review should:

a) be documented;

b) become the basis for making decisions regarding the implementation of the OSH management system; and

c) provide a baseline from which continual improvement of the organisation's OSH management system can be measured.

System planning, development and implementation

The purpose of planning should be to create an OSH management system that supports:

a) as the minimum, compliance with national laws and regulations;

b) the elements of the organisation's OSH management system; and

c) continual improvement in OSH performance.

Arrangements should be made for adequate and appropriate OSH planning, based on the results of the initial review, subsequent reviews or other available data. These planning arrangements should contribute to the protection of safety and health at work, as well as achieving organisational goals.

This stage requires organisations to

- compliance with health and safety related laws
- establish health and safety targets
- manage change within the organisation to address the improving safety culture
- identification of workplace hazards including the assessment of risk and determine how these risks can be eliminated or further controlled
- set standards against which performance can be measured
- develop and introduce safe systems of work that reduce risk, this encompassing the design tasks, processes, equipment to be utilised
- provision of protective equipment where risks can not be adequately controlled by other means
- development of emergency preparedness procedures
- establish contractor compliance and co-operation

Evaluation – Performance Monitoring and Measurement

As with any other business function measuring performance is a crucial step and needs to be executed with care to achieve accurate reporting.

Procedures to monitor, measure and record OSH performance on a regular basis should be developed, established and periodically reviewed. Responsibility, accountability and authority for monitoring at different levels in the management structure should be allocated.

The selection of performance indicators should be according to the size and nature of activity of the organisation and the OSH objectives.

Both qualitative and quantitative measures appropriate to the needs of the organisation should be considered. These should:

a) be based on the organisation's identified hazards and risks, the commitments in the OSH policy and the OSH objectives; and

b) support the organisation's evaluation process, including the management review.

Performance monitoring and measurement should:

a) be used as a means of determining the extent to which OSH policy and objectives are being implemented and risks are controlled;

b) include both active and reactive monitoring, and not be based only upon work related injury, ill health, disease and incident statistics; and

c) be recorded.

Active monitoring

This should contain the elements necessary to have a proactive system and should include:

a) monitoring of the achievement of specific plans, established performance criteria and objectives;

b) the systematic inspection of work systems, premises, plant and equipment;

c) surveillance of the working environment, including work organisation;

d) surveillance of workers' health, where appropriate, through suitable medical monitoring or follow-up of workers for early detection of signs and symptoms of harm to health in order to determine the effectiveness of prevention and control measures; and

e) compliance with applicable national laws and regulations, collective agreements and other commitments on OSH to which the *organisation* subscribes.

Reactive monitoring

Should include the identification, reporting and investigation of:

a) work-related injuries, ill health (including monitoring of aggregate sickness absence records), diseases and incidents;

b) other losses, such as damage to property;

c) deficient safety and health performance, and OSH management system failures; and

d) workers' rehabilitation and health-restoration programmes.

Audit

Arrangements to conduct periodic audits are to be established in order to determine whether the OSH management system and its elements are in place, adequate, and effective in protecting the safety and health of workers and preventing incidents.

An audit policy and programme should be developed, which includes a designation of auditor competency, the audit scope, the frequency of audits, audit methodology and reporting.

Management reviews

Management reviews should:

a) evaluate the overall strategy of the OSH management system to determine whether it meets planned performance objectives;

b) evaluate the OSH management system's ability to meet the overall needs of the organisation and its stakeholders, including its workers and the regulatory authorities;

c) evaluate the need for changes to the OSH management system, including OSH policy and objectives;

d) identify what action is necessary to remedy any deficiencies in a timely manner, including adaptations of other aspects of the organisation's management structure and performance measurement;

e) provide the feedback direction, including the determination of priorities, for meaningful planning and continual improvement;

f) evaluate progress towards the organisation's OSH objectives and corrective action activities; and

g) evaluate the effectiveness of follow-up actions from earlier management reviews.

The frequency and scope of periodic reviews of the OSH management system by the employer or the most senior accountable person should be defined according to the organisation's needs and conditions.

The management review should consider:

- the results of work-related injuries, ill health, diseases and incident investigations; performance monitoring and measurement; and audit activities; and
- additional internal and external inputs as well as changes, including organisational changes, that could affect the OSH management system.

The findings of the management review should be recorded and formally communicated to:

a) the persons responsible for the relevant element(s) of the OSH management system so that they may take appropriate action; and

b) the safety and health committee, workers and their representatives as appropriate.

Action for improvement

Arrangements should be established and maintained for preventive and corrective action resulting from OSH management system performance monitoring and measurement, OSH management system audits and management reviews. These arrangements should include:

a) identifying and analysing the root causes of any non-conformities with relevant OSH regulations and/or OSH management systems arrangements; and

b) initiating, planning, implementing, checking the effectiveness of and documenting corrective and preventive action, including changes to the OSH management system itself.

When the evaluation of the OSH management system or other sources show that preventive and protective measures for hazards and risks are inadequate or likely to become inadequate, the measures should be addressed according to the recognized hierarchy of prevention and control measures, and completed and documented, as appropriate and in a timely manner.

Continual improvement

Arrangements should be established and maintained for the continual improvement of the relevant elements of the OSH management system and the system as a whole. These arrangements should take into account:

a) the OSH objectives of the organisation;

b) the results of hazard and risk identifications and assessments;

c) the results of performance monitoring and measurements;

d) the investigation of work-related injuries, diseases, ill health and incidents, and the results and recommendations of audits;

e) the outcomes of the management review;

f) the recommendations for improvement from all members of the organisation, including the safety and health committee, where it exists;

g) changes in national laws and regulations, voluntary programmes and collective agreements;

h) new relevant information; and

i) the results of health protection and promotion programmes.

The safety and health processes and performance of the organisation should be compared with others in order to improve health and safety performance.

OHSAS 18001 (BSI)

The original British Standard, BS 8800:1996 Guide to Occupational Health and Safety Management Systems, while fulfilling its original purpose of providing guidance, was found lacking by a number of organisations that wished to certify their safety management systems, very much like those for quality and environmental management. As a result, the British Standards Institution, in conjunction with six certification bodies and eight standards bodies, produced a standard for a safety management system that was capable of supporting certification.

The international bodies working on the standard were:

1. UK
2. Ireland
3. Spain
4. Australia
5. Singapore
6. South Africa
7. Malaysia and

8. Japan

OHSAS 18001 was developed to be compatible with the two other certification standards, that of BS EN ISO 9001:1994 for quality management and BS EN ISO 14001 for environmental management. As a result, organisations seeking certification to one or both of these two management schemes could usefully encompass their safety management scheme in with these to complete the whole quality, environment, safety and health (QUENSH) management package.

The principles of OHSAS 18001 are similar to those of HSG65 (planning, implementing, reviewing and modifying) and so the HSE Guidance HSG 65 could be studied as a basis for preparation of a certifiable safety management system under this standard. This is discussed later in this Element.

The British Standards Institution (BSI) has also prepared a companion standard, BS EN ISO 18002:2000 Occupational Health and Safety Management Systems. Guidelines for the implementation of OHSAS 18001, which enables organisations to prepare their system for compliance with the OHSAS 18001, and on to certification of that compliance.

The OH&S Policy

An OH&S policy establishes an overall sense of direction within an organisation and sets out the principles of action for OH&S. As with other 'management Systems' the OH&S policy should be properly documented, implemented and of course maintained. It should also be consistent with the organisations overall business policies, including those policies for other managerial disciplines i.e. quality and environmental management.

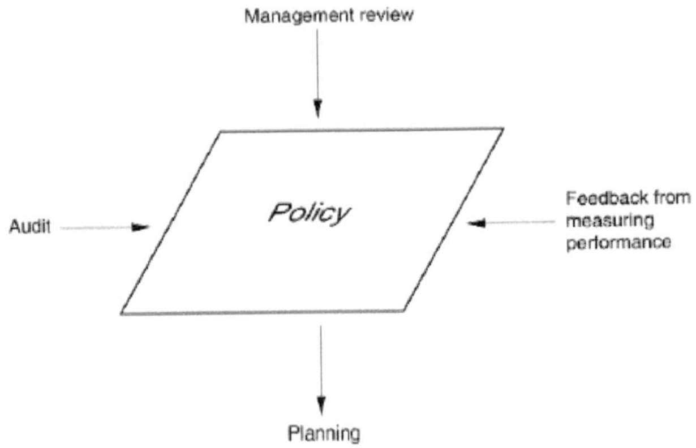

The policy shall:

a) be appropriate to the nature and scale of the organization's OH&S risks;

b) include a commitment to continual improvement;

c) include a commitment to at least comply with current applicable OH&S legislation and with other requirements to which the organization subscribes;

d) be documented, implemented and maintained;

e) be communicated to all employees with the intent that employees are made aware of their individual OH&S obligations;

f) be available to interested parties; and

g) be reviewed periodically to ensure that it remains relevant and appropriate to the organization.

Planning

The organization shall establish and maintain procedures for the ongoing identification of hazards, the assessment of risks, and the implementation of necessary control measures.

These shall include:

- routine and non-routine activities;
- activities of all personnel having access to the workplace (including subcontractors and visitors);
- facilities at the workplace, whether provided by the organization or others.

The organization shall ensure that the results of these assessments and the effects of these controls are considered when setting its OH&S objectives. The organization shall document and keep this information up to date.

The organization's methodology for hazard identification and risk assessment shall:

- be defined with respect to its scope, nature and timing to ensure it is proactive rather than reactive;
- provide for the classification of risks and identification of those that are to be eliminated or controlled by measures as defined in the standard;
- be consistent with operating experience and the capabilities of risk control measures employed;

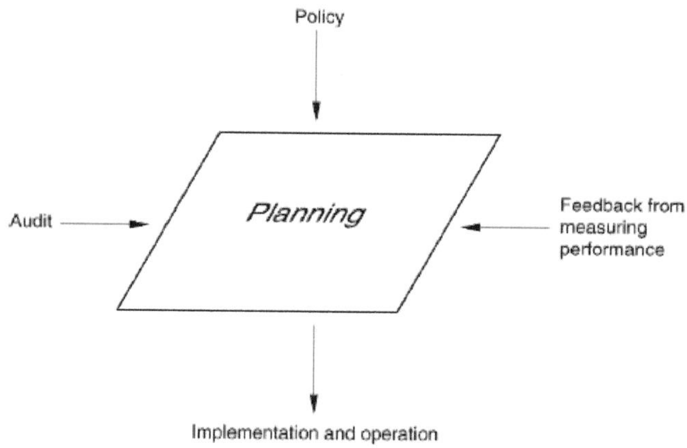

- provide input into the determination of facility requirements, identification of training needs and/or development of operational controls;
- provide for the monitoring of required actions to ensure both the effectiveness and timeliness of their implementation.

OHSAS 18002 does provide further guidance on hazard identification, risk assessment and the control of risk.

Implementation and Operation

The roles, responsibilities and authorities of personnel who manage, perform and verify activities having an effect on the OH&S risks of the organization's activities, facilities and processes, shall be defined, documented and communicated in order to facilitate OH&S management.

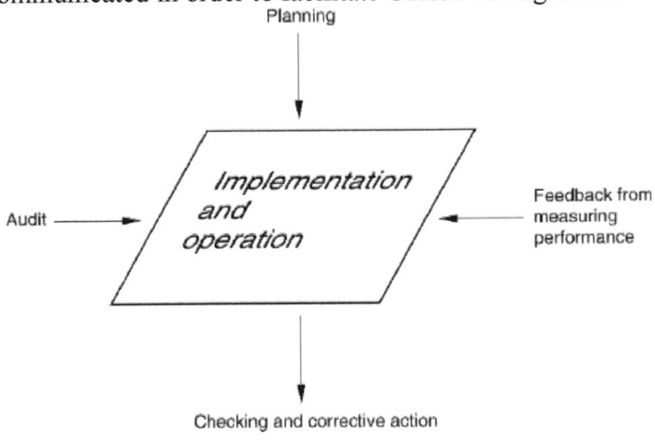

Control

Ultimate responsibility for occupational health and safety rests with top management. The organization shall appoint a member of top management (e.g. in a large organization, a Board or executive committee member) with particular responsibility for ensuring that the OH&S management system is properly implemented and performing to requirements in all locations and spheres of operation within the organization.

The organization's management appointee shall have defined roles, responsibilities and authority for:

a) ensuring that OH&S management system requirements are established, implemented and maintained in accordance with this OHSAS specification;

b) ensuring that reports on the performance of the OH&S management system are presented to top management for review and as a basis for improvement of the OH&S management system.

All those with management responsibility shall demonstrate their commitment to the continual improvement of OH&S performance.

Competence

Personnel shall be competent to perform tasks that may impact on OH&S in the workplace. Competence shall be defined in terms of appropriate education, training and / or experience.

The organization shall establish and maintain procedures to ensure that its employees working at each relevant function and level are aware of:

- the importance of conformance to the OH&S policy and procedures, and to the requirements of the OH&S management system;
- the OH&S consequences, actual or potential, of their work activities and the OH&S benefits of improved personal performance;
- their roles and responsibilities in achieving conformance to the OH&S policy and procedures and to the requirements of the OH&S management system,

including emergency preparedness and response requirements;

- the potential consequences of departure from specified operating procedures.

Training procedures shall take into account differing levels of:

- responsibility, ability and literacy; and
- risk.

Consultation and Communication

The organization shall have procedures for ensuring that pertinent OH&S information is communicated to and from employees and other interested parties.

Employee involvement and consultation arrangements shall be documented and interested parties informed.

Employees shall be:

a) involved in the development and review of policies and procedures to manage risks;

b) consulted where there are any changes that affect workplace health and safety;

c) represented on health and safety matters; and

d) informed as to who is their employee OH&S representative(s) and specified management appointee

Operational Control

The organization shall identify those operations and activities that are associated with identified risks where control measures need to be applied. The organization shall plan these activities, including maintenance, in order to ensure that they are carried out under specified conditions by:

a) establishing and maintaining documented procedures to cover situations where their absence could lead to deviations from the OH&S policy and the objectives;

b) stipulating operating criteria in the procedures;

c) establishing and maintaining procedures related to the identified OH&S risks of goods, equipment and services purchased and/or used by the organization and communicating relevant procedures and requirements to suppliers and contractors;

d) establishing and maintaining procedures for the design of workplace, process, installations, machinery, operating procedures and work organization, including their adaptation to human capabilities, in order to eliminate or reduce OH&S risks at their source.

Checking and Corrective Action

The organization shall establish and maintain procedures to monitor and measure OH&S performance on a regular basis. These procedures shall provide for:

- both qualitative and quantitative measures, appropriate to the needs of the organization;
- monitoring of the extent to which the organization's OH&S objectives are met;
- proactive measures of performance that monitor compliance with the OH&S management programme, operational criteria and applicable legislation and regulatory requirements;
- reactive measures of performance to monitor accidents, ill health, incidents (including near-misses) and other historical evidence of deficient OH&S performance;
- recording of data and results of monitoring and measurement sufficient to facilitate subsequent corrective and preventative action analysis.

•

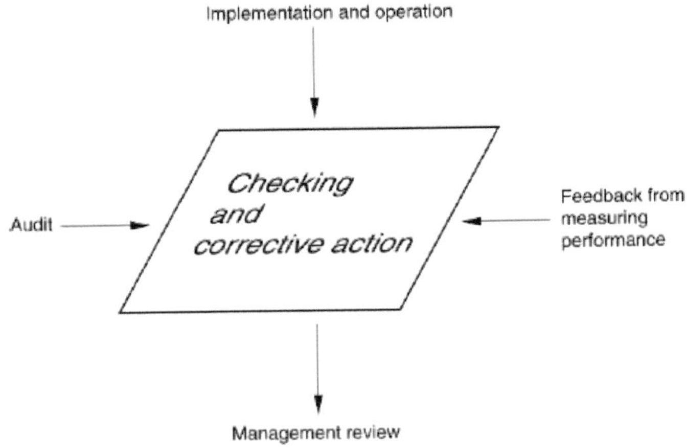

If monitoring equipment is required for performance measurement and monitoring, the organization shall establish and maintain procedures for the calibration and maintenance of such equipment. Records of calibration and maintenance activities and results shall be retained.

Management Review

The organization's top management shall, at intervals that it determines, review the OH&S management system, to ensure its continuing suitability, adequacy and effectiveness. The management review process shall ensure that the necessary information is collected to allow management to carry out this evaluation. This review shall be documented.

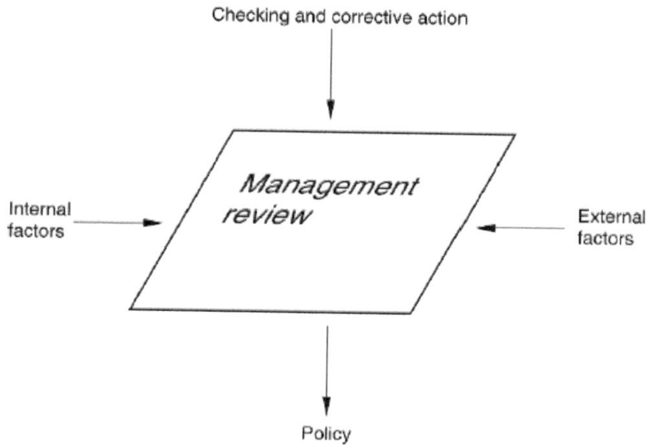

The management review shall address the possible need for changes to policy, objectives and other elements of the OH&S management system, in the light of OH&S management system audit results, changing circumstances and the commitment to continual improvement.

The management system should be reviewed on a regular basis to ensure its continued suitability and effectiveness for the organisation to deliver its commitment as well as continual improvements.

Should any changes to the system be introduced then these will need to be communicated to all concerned as soon as is practicable.

HSG 65 (HSE)

Policy

Effective health and safety policies set a clear direction for the organisation to follow. They contribute to all aspects of business performance as part of a demonstrable commitment to continuous improvement. Responsibilities to people and the environment are met in ways which fulfil the spirit and letter of the law. Stakeholders' expectations in the activity (whether they are shareholders, employees, or their representatives, customers or society at large) are satisfied. There are cost-effective approaches to preserving and developing physical and human resources, which reduce financial losses and liabilities.

Organising

An effective management structure and arrangements are in place for delivering the policy. All staff are motivated and empowered to work safely and to protect their long-term health, not simply to avoid accidents. The arrangements are:

- underpinned by effective staff involvement and participation; and
- sustained by effective communication and the promotion of competence which allows all employees and their representatives to make a responsible and informed contribution to the health and safety effort.

There is a shared common understanding of the organisation's vision, values and beliefs.

A positive health and safety culture is fostered by the visible and active leadership of senior managers.

Planning

There is a planned and systematic approach to implementing the health and safety policy through an effective health and safety management system. The aim is to minimise risks. Risk assessment methods are used to decide on priorities and to set objectives for eliminating hazards and reducing risks.
Wherever possible, risks are eliminated through selection and design of facilities, equipment and processes. If risks cannot be eliminated, they are minimised by the use of physical controls or, as a last resort, through systems of work and personal protective equipment. Performance standards are established and used for measuring achievement. Specific actions to promote a positive health and safety culture are identified.

Measuring performance

Performance is measured against agreed standards to reveal when and where improvement is needed. Active self-monitoring reveals how effectively the health and safety management system is functioning. This looks at both hardware (premises, plant and substances) and software (people, procedures and systems) including individual behaviour and performance.

If controls fail, reactive monitoring discovers why by investigating accidents, ill health or incidents which could cause harm or loss.

The objectives of active and reactive monitoring are:

- to determine the immediate causes of sub-standard performance; and
- to identify the underlying causes and the implications for the design and operation of the health and safety management system.

Longer-term objectives are also monitored.

Auditing and reviewing performance

The organisation learns from *all* relevant experience and applies the lessons. There is a systematic review of performance based on data from monitoring and from independent audits of the whole health and safety management system.

There is a strong commitment to continuous improvement involving the constant development of policies, systems and techniques of risk control. Performance is assessed by:

- internal reference to key performance indicators; and
- external comparison with the performance of business competitors and best practice, irrespective of employment sector.

APPENDEX – I: BIBLIOGRAPHY

Publishing Organisation	Year	Title
ILO	1981	C 155 – Occupational Safety and Health Convention
ILO	1981	R164 -Occupational Safety and Health Recommendation
ILO	2001	ILO-OSH –Guidelines on Occupational Safety and Health Management Systems
ILO	2005	ILOLEX – ILO Database of International Law
BSI	1999	OHSAS - 18000 Occupational Health and Safety Assessment Series
BSI	1999	OHSAS 18001 – Occupational Health and Safety Management Systems Specification
BSI	2002	OHSAS 18002 – Occupational health and Safety Management Systems – Guidelines For The Implementation of HSAS 18001
HSE	2003	HSG 65 – Successful Health and Safety Management
HSE	1997	HSG 96 - The Cost of Accidents at Work
HSE	2004	L 24 – ACoP and Guidance Workplace (Health, Safety and Welfare) Regulations 1992 (as amended)

APPENDEX – II: ABOUT AUTHOR

Peter C Barnett-Schuster is a Chartered Safety and Health Practitioner and registered consultant with considerable years experience in the health and safety field. His career has taken him throughout the United Kingdom, Europe and the Middle East where he has specialized in training and consultancy services to the construction industry.

Presently the Vice Chairman for the Institution of Safety and Health (IOSH) Middle East Branch and an active member of other committee's within the region, he is well respected and often a guest speaker at seminars around the United Arab Emirates and adjoining countries.

His aim is to raise awareness through education and in turn assist in the un-necessary fatalities, major injuries and accidents from occurring in the workplace.

Peter intends to develop a series of health and safety books in a simplified context which everyone will understand and be able to digest and act upon.

APPENDEX – III: INDEX

A

C

E

F

G

H

I

NOTES

www.ingramcontent.com/pod-product-compliance
Ingram Content Group UK Ltd.
Pitfield, Milton Keynes, MK11 3LW, UK
UKHW041141300926
15471UKWH00011B/89

9 780615 214900